The Wild Side of Pet Ferrets

Jo Waters

Raintree

Chicago, Illinois

For information, address the publisher:
Raintree, 100 N. LaSalle, Suite 1200, Chicago, IL 60602

Customer Service: 888-363-4266
Visit our website at www.raintreelibrary.com

Printed and bound in China by South China Printing Company.
08 07 06 05 04
10 9 8 7 6 5 4 3 2 1

Library of Congress Cataloging-in-Publication Data
Waters, Jo.
  The wild side of pet ferrets / Jo Waters.
      p. cm. -- (The wild side of pets)
Includes bibliographical references and index.
  ISBN 1-4109-1019-9 (lib. bdg.) -- ISBN 1-4109-1159-4 (pbk.)
  1. Ferrets as pets--Juvenile literature. 2. Ferrets--Juvenile
literature. [1. Ferrets as pets. 2. Ferrets. 3. Pets.] I. Title.
II. Series: Waters, Jo. Wild side of pets.
SF459.F47W38 2005
636.976'628--dc22

                              2003024751

Acknowledgments
The author and publisher would like to thank the following for permission to reproduce photographs: pp. 4, 6 A. Williams/NHPA; p. 5 top M. Leach/Oxford Scientific Films; p. 5 bot J. Wegner/Bruce Coleman Collection; pp. 7, 11, 13, 15, 29 Dave Bradford; pp. 9, 17, 19, 21, 23, 25 Tudor photography/Harcourt Education Ltd; p. 10 J. Cancalois/Nature Picture Library; p. 12 D. Fox/Oxford Scientific Films; pp. 14, 24 D. Kjaer/Nature Picture Library; p. 16 Silvestris/ FLPA; p. 20 Ecoscene; p. 22 NHPA; p. 26 E. Degginger/Oxford Scientific Films; p. 27 M. Barton/Nature Picture Library; p. 28 W. Shattil & B. Rozinski/Oxford Scientific Films.

Cover photograph of a pet ferret, reproduced with permission of Paul Beard/Sue Redman.
Inset cover photograph of a European polecat reproduced with permission of Mark Bowler/ NHPA.

The publishers would like to thank Michaela Miller for her assistance in the preparation of this book.

Every effort has been made to contact copyright holders of any material reproduced in this book. Any omissions will be rectified in subsequent printings if notice is given to the publishers.

Contents

Some words are shown in bold, **like this.** You can find out what they mean by looking in the Glossary.

Was Your Pet Once Wild?

Did you know that your pet ferret is related to wild animals? Finding out more about your ferret's wild **ancestors** will help you give it a better life.

Polecats are the wild animals that are the most closely related to ferrets. Like polecats, ferrets are small **mammals** with long, slender bodies, thick fur coats, sharp claws, and strong paws.

Bad smell!

*The **Latin** name for polecats is Mustela putorius. Mustela means "mouse-catcher" and putor means "bad smell." Polecats can make a nasty smell to frighten off **predators.***

Domestic ferrets were first bred from wild polecats like this one.

4

Ferrets can be great pets. They need a lot of looking after but this can be fun. Ferrets should be brought into homes with young children very carefully. Ferrets can be very lively and may bite if frightened or hurt.

Ferrets and polecats are very similar.

How ferrets became pets

We know that the ancient Egyptians and Romans kept ferrets as pets. Ferrets came from polecats. They were used to help people hunt small animals like rabbits.

Types of Ferret

There are three main types of polecat: European polecats, Steppe polecats, and marbled polecats. They all look similar.

Marbled polecats have yellow and white marks on a red and brown back. Their belly is black and they have a white stripe down their forehead. European and steppe polecats are larger and are usually a brown color.

Polecats usually have markings across their eyes that look like a "robber-mask," like this European polecat.

Ferret relatives

Polecats and ferrets are part of the mustelid family. This includes weasels, otters, pine martens, and badgers.

Pet ferrets can be black, brown, multi-colored, and even white.

Special markings

Ferrets can have different coat colors or markings. These include all-white **albino,** "polecat," silver, and sandy.

Ferrets need lots of care and handling to keep them tame and friendly.

Pink eyes

Albino ferrets are ferrets with no color in their skin, eyes, or fur. They have red or pink eyes. There is nothing wrong with them.

Where Are Ferrets From?

Wild polecats live all over the world. The European polecat lives in Europe, northern and western Asia, and in North Africa. The Steppe polecat lives in central Asia. The marbled polecat is found in southern Europe and western China.

Survival

In the wild, only the fittest, strongest polecats survive, so polecats usually look very alike. Different looking polecats may not be as good at hunting, because they are more easily spotted by **prey.**

This map shows where polecats, ferrets, and their relatives live.

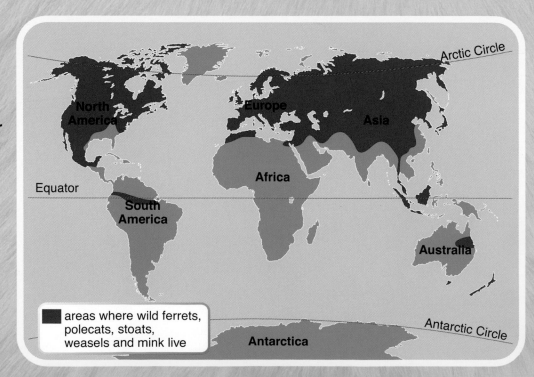

Arctic Circle

North America

Europe

Asia

Africa

Equator

South America

Australia

areas where wild ferrets, polecats, stoats, weasels and mink live

Antarctica

Antarctic Circle

8

Sometimes ferrets lose their home. You might be able to adopt a ferret from an animal shelter, like this one.

Keeping ferrets is very popular in the United States, England, Europe, and Australia. When you get a ferret, make sure it comes from a good home. You can ask a local veterinarian to recommend a good place. Check that your vet treats ferrets before buying your new pet, because some do not.

Choosing a ferret

Ferrets should have bright, clear eyes and nose, glossy fur, and a clean bottom. They should have plenty of energy.

Ferret habitats

In the wild, polecats can live in all kinds of **habitats.** This includes forests, mountains, farms, deserts, grasslands, and other bushy areas. You can often find them anywhere that their **prey** lives.

Polecats have adapted to live in different places. In Great Britain, polecats are found in forests and rocky places in Wales and Scotland.

Camouflage
European polecats have dark fur with lighter tips on their ears and sometimes on their nose and belly. Their dark-colored fur helps them blend in with their surroundings.

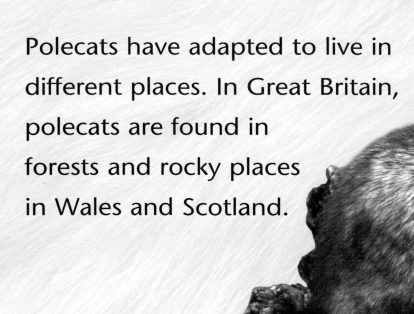

The European polecat can be found all over Europe.

Curious creatures

Just like polecats, pet ferrets like to roam and explore. Ferrets are very curious, so keep a careful eye on them. They can get stuck or injured in washing machines, dishwashers, garbage cans, and cupboards.

Ferrets need a roomy cage with a solid floor for their main home.

Ferrets love human company and should be with people for at least four hours every day. Ferret cages can be kept indoors or in a sheltered outdoor area. Their home should be cleaned once a week and their litter area cleaned every day.

Ferret Anatomy

All polecats and ferrets have the same basic **anatomy,** or body parts. Polecats have large front paws for digging and a wide tail. They all have thick, protective fur. They have strong, sharp teeth designed for catching and eating small **mammals** and insects.

Agility

Polecats are all slim and **agile,** with a flexible spine and short legs. This means they can bend and turn incredibly tightly.

Polecats have five toes and claws. Claws are useful for gripping the ground when running and clinging when climbing.

Ferrets are very similar to their wild relatives. They also have short legs, a flexible back, claws, and sharp teeth. Pet ferrets use their paws and claws just like polecats. Ferrets may try to dig or burrow under things. Do not leave gaps where ferrets can get under carpets or flooring.

Ferrets usually weigh up to two pounds (one kilogram). Male ferrets can be over twice the size of female ferrets.

This drawing shows the skeleton of a ferret.

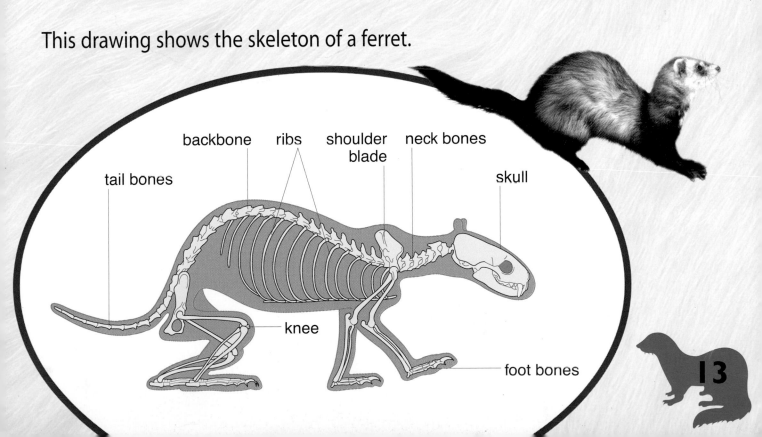

backbone   ribs   shoulder   neck bones
                    blade

tail bones   skull

knee

foot bones

Senses

Polecats have very sharp senses of smell and hearing. They use smell to hunt and find their **prey.**

Polecats can move their ears to hear in different directions without having to turn their heads. They can hear higher sounds than humans can.

Polecats do not use their eyes much. They live in small, dark spaces such as tunnels so they are very short-sighted.

Polecats have "eyebrows" made of eight whiskers above each eye.

Just like polecats, pet ferrets do not see very well. They can sometimes fall from high places because they have not seen the drop. Make sure your ferret cannot climb up to dangerous places.

Ferrets can hear much higher sounds than us.

Sensitive animals

Pet ferrets have kept their wild **ancestor's** amazing sense of smell. A ferret's sense of smell is much better than ours.

Ferrets also have excellent hearing. It can be damaged by loud bangs, TV noise, or music, which they would not hear in the wild. Make sure you keep your ferret somewhere quiet.

Movement

Polecats have very thin bodies.

Polecats use the muscles in their backs to jump and climb and also to run. They use their long back and short legs to keep low to the ground. They can make themselves very long and thin to squeeze through tight tunnels easily.

Swimming

Most polecats can swim very well. They do not really enjoy it and will avoid water if possible. They paddle with all four feet while keeping their noses clear of the water to breathe.

Pet ferrets move in the same way as polecats. It is important that they have lots of space to exercise in.

Handling ferrets

Ferrets are quite wriggly so you must pick them up the right way. Use both hands, one around the ferret's back legs and the other around its shoulders. Hold the ferret firmly and keep it close to your body.

Your ferret can be trained to walk on a leash and **harness** if you want to take it outside. Ferrets should not be taken far, but it is exciting for them to be able to explore a new area.

With a harness and leash, you can take your ferret for a walk.

17

What Do Ferrets Eat?

Polecats eat meat. They hunt small **mammals** and insects for food. They cannot eat fruit, vegetables, or grains very well.

Polecats need to eat every day. Polecats also need to drink plenty of water to stay healthy.

Fat and thin

Polecats will put on about half their body weight in fat during the summer. There is less food around in winter, so polecats need this body fat to keep them alive.

This is how polecats fit into a **food chain.**

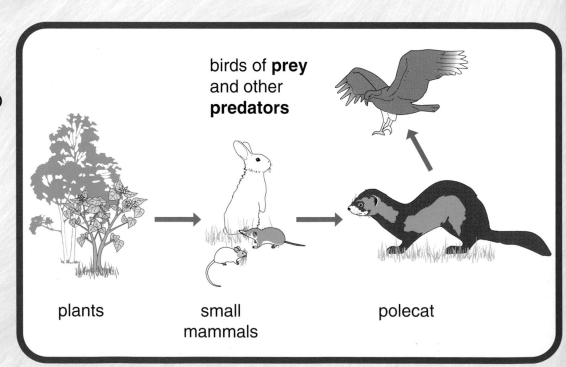

birds of **prey** and other **predators**

plants   small mammals   polecat

18

Dried ferret foods are easy to feed and are also crunchy, so they keep your ferret's teeth clean.

Just like polecats, ferrets eat meat. Giving them fruit, vegetables, or grains can make them ill.

Small amounts of scrambled egg or boiled fish make good ferret treats. Never give your ferret chocolate, licorice, or onions. They can be poisonous. Milk can upset their stomachs.

Drinking

*Your ferret needs lots of fresh, clean water. A **dispenser bottle** is good because the ferrets cannot tip it over or make the water dirty.*

Hunting and Playing

Wild polecats hunt small **mammals,** like rabbits, by going down their burrows or into their nests.

All young polecats play. This is how they learn their hunting and **social** skills. They wrestle with their brothers and sisters to see who is strongest. They play pouncing and chase games to practice hunting.

Rough and tumble

Polecat play is often very rough. Young polecats will bite their playmates and drag them around. They are not usually harmed because they have very tough skins.

Polecats hunt and eat alone.

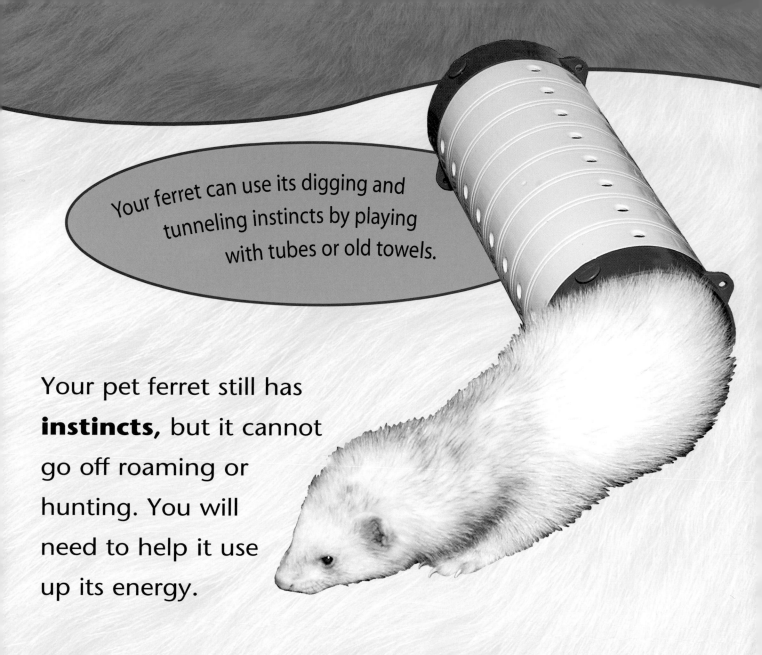

Your ferret can use its digging and tunneling instincts by playing with tubes or old towels.

Your pet ferret still has **instincts,** but it cannot go off roaming or hunting. You will need to help it use up its energy.

Playing is good exercise. You can get special ferret squeaky toys. Ferrets see the toy as **prey.** They use their wild instincts to chase and catch.

Biting

Ferrets will sometimes nip you in play. Our skin is not as tough as ferret skin so a nip can hurt. Remember that your pet is not being mean. Nipping is natural ferret behavior.

21

Living in Groups

Polecats prefer to live alone. They group together only when the male and female **mate.** Then the female will look after her babies until they are ready to go off by themselves.

Fighting

If males meet, they will probably fight. Each polecat will attack the back of the other polecat's neck. They also attack sideways or dance around, making a clacking noise. This is to scare the other polecat away. Polecats may scream or hiss as well, as a sign of fear or pain.

Polecats make *noises* when frightened or *angry.*

22

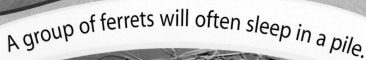
A group of ferrets will often sleep in a pile.

Ferrets are different from polecats. Pet ferrets never grow out of their baby behavior so they like living in a "family." A ferret should always be kept with other ferrets.

Body language

Ferrets use their bodies to **communicate.** *Trembling can mean a ferret is scared. Licking another ferret's ears can mean "sorry." If an adult ferret gets fed up with babies annoying it, it may pin them down and wash their ears to say "leave me alone!"*

Sleeping

Polecats need to sleep a lot, sometimes up to eighteen hours a day.

Polecats are active at night and sleep during the day.

Polecat dens

Polecats sleep in burrows or dens. They dig these out of the ground or under tree roots, or they find a space like a hidden cave or shelter. They will also take over other animals' burrows.

This polecat has a den underground.

Pet ferrets sleep during the day and night in naps, and wake up for periods in between. Ferrets are usually awake when you get up to go to school in the morning and when you get home in the evening.

Bedding

Ferrets should have a cozy bed to sleep in. Shredded paper and old blankets make good bedding. They also like to sleep in drawers or under sofas.

Ferrets like dark, safe places to sleep. Your ferret needs a cage like a polecat would have a burrow or a den.

Ferrets love to burrow in old blankets.

Life Cycle of Ferrets

A female polecat is called a jill and a male polecat is called a hob. Young polecats are called kits.

Polecats are **pregnant** for around six weeks. They have their kits in their burrows. Polecats can have up to twelve kits, but they normally have between six and nine.

Polecats will live for as long as five years in the wild.

The kits get a thick coat by about 4 to 5 weeks.

26

Ferret kits usually stay in the nest for the first 3 to 4 weeks.

Pet ferrets can live twice as long as wild polecats. This is because they do not need to hunt for food to survive.

Ferrets are born completely helpless, bald, and blind. Kits can usually leave the litter at about eight weeks. This is a good time to get a ferret since it will adjust quickly to you being its new family.

Neutering

Both male and female ferrets should be **neutered** to stop them from having babies. This can also make them less aggressive.

Common Problems

Polecats and their other wild relatives, like weasels and otters, may be **endangered** by **pollution** and damage to their **habitats.**

In danger!

*Some polecats are killed by humans. There used to be thousands of black-footed ferrets living wild in the United States. Farmers killed them because they thought polecats were pests. In order to stop them from dying out completely, scientists have been **breeding** black-footed ferrets. They then release the ferrets back into the wild in Wyoming.*

The black-footed ferret lives wild in the U.S.

These are some common ferret problems.

tooth decay

ear mites

fleas and ticks

worms

Itches and sneezes

Ferrets can get **parasites.** Smelly wax in their ears can mean they have ear **mites.** A veterinarian will recommend a regular treatment for your ferrets.

Ferrets can catch human coughs and colds. If you have a cold, do not handle your ferret. Ferrets should also be **vaccinated** against diseases. Ask your vet which vaccinations your ferrets should have.

Find Out for Yourself

A good owner will always want to learn more about keeping a pet ferret. To find out more information about ferrets, you can look in other books and on the Internet.

Books to read

Gelman, Amy. *My Pet Ferrets.* Minneapolis, Minn.: Lerner Publications, 2000.

McNicholas, June. *Ferrets.* Chicago: Heinemann Library, 2002.

Using the Internet

Explore the Internet to find out about ferrets. Websites can change, so if one of the links below no longer works, do not worry. Use a search engine, such as *www.google.com* or *www.internet4kids.com*. You could try searching with the keywords "ferret," "pet," and "polecat."

Glossary

agile able to move fast and easily

albino completely without color. Albino ferrets have white fur and pink eyes.

anatomy how the body is made

ancestor animals in the past from which today's animals are descended

breed animals mate and have babies

communicate to make yourself understood

dispenser bottle special bottle with a pipe for drinking from

domestic animal that has been bred to be tame

endangered in danger of dying or being killed

food chain links between different animals that feed on each other and on plants

habitat where an animal or plant lives

harness straps that wrap around the body of a ferret

instinct natural behavior that an animal is born with

mammal animal with warm blood who gives birth to live babies

mate come together with another animal to make babies

mite small parasite that lives on another animal and sucks its blood

neutered animal that has had an operation so it cannot have babies

parasite tiny animal that lives in or on another animal and feeds off it

pollution making the environment dirty with waste and poisonous chemicals

predator animal that hunts and eats other animals

pregnant to have a baby growing inside

prey animals that are hunted and eaten by other animals

social likes company and living in groups

vaccinate give injections to protect against diseases

31

Index